Introduction

This book was designed to be a creative tool to aid in the teaching and understanding of related facts and their inverse relationship. This book has been successfully used in my own and many of my colleague's classrooms.

Teaching keys to remember and emphasize:

- In addition sentences, the big number (*daddy*) must always go last.
- In subtraction sentences, the big number (*daddy*) must always come first.
- Mommy (*middle* number) and baby (*little* number) can switch places with each other in either kind of number sentence.
- Either mommy or baby can start the first row because they can switch places anytime.
- To be related, the numbers in a number sentence must be the same numbers as in another number sentence.

Here is the story in non-rhyming form to be used with manipulatives:

One day there was a baby, a mommy and a daddy. Baby liked to play with mommy while daddy was doing his chores. Mommy liked to play with baby when daddy was at work. But when daddy comes home from work he says, "That's enough! Now daddy gets to play with baby while mommy goes away." But sometimes, daddy likes to spend time with mommy while baby takes a nap.

(Remember that either baby or mommy can start the story because they can change places with each other in a number sentence.)

***** This book can also be used to aid in the understanding of inverse relationships in multiplication and division.**

THIS BOOK IS DEDICATED TO ALL OF MY FIRST GRADE STUDENTS BOTH PAST AND PRESENT!

*A special thank you goes out to my teammate,
Alice Thomas for giving me the idea of using a family to
teach related facts.

First Edition

Library of Congress Cataloging-in-Publication Data
Turley, Sandy
"The Fact Family"
by Sandy Turley; illustrations by Holly Parson Nielsen

Summary: A creative, motivating rhyming story about a family named Fact that teaches the inverse relationship between the numbers in a number sentence.

Library of Congress Control Number: 2006901903
ISBN 10: 09778548-1-7
ISBN 13: 9-780977-854813

1. Families--Juvenile Fiction 2. Number relationships--Juvenile Fiction
3. Math Facts--Juvenile Fiction 4. Addition and subtraction--Juvenile Fiction
5. [Fact Families] Title.

The Fact Family

A Teaching Rhyme About Inverse Number Relationships

Written by Sandy Turley
Illustrated by Holly Parson Nielsen

Once there was a family
with the last name of Fact.
Who loved being together,
everyday to be exact.

This family had a story
that they loved to tell.
To their goods friends,
they liked very well.

Important family secrets
this story helped to show.
It had been told for years,
since very long ago.

When they told their story,
a big house they drew.
It had five rows of windows,
fifteen in all to view.

Three windows in each row,
on the house you'd see,
So each family member
in five windows could be.

So here is the story
this little family told.
Listen very closely,
as secrets will unfold.

One day there was a baby,
a mommy and daddy too.
Mommy liked to play with baby,
while daddy his chores did do.

The baby and the mommy
also liked to play,
When daddy was at work,
each and every day.

When daddy got home from work
"That's enough!, he would say.
"It's my turn to play with baby
while mommy goes away."

But sometimes after work,
mommy he wanted to see,
So when baby took a nap,
together they would be.

Looking at the Fact family house,
these secrets you might have found,
That mommy and little baby
like to move around.

You also might have noticed,
the beginning or the end,
Are the only windows
where daddy time can spend.

This little family story
about the related facts,
Actually involves three numbers,
which are a link to be exact.

4 - 3 = 1

In a number sentence,
three numbers we usually find.
Each member of the Fact family,
one number is assigned.

Daddy is the largest number,
mommy is the middle.
Baby is the number that's left,
the number which is little.

$$4 - 3 = 1$$

The story begins at the top,
where the numbers we will see,
Baby, mommy and daddy
in a window each will be.

3
4

Mommy starts the first row,
in the first window she does sit.
Then comes the plus
which you should not forget.

Then it's baby's turn,
he sits in window two.
The equal sign comes next,
the first row's almost through.

In the last window,
Daddy you will find.
Many windows still are left
that need to be assigned.

On the second row,
mommy and baby move around.
They do this very quietly,
making not a sound.

Baby now comes first,
then comes the plus.
Next comes mommy,
now equal is a must.

Daddy is at the end,
this finishes the second row.
There are two more rows to tend,
before it's time to go.

The third row's a little different,
in the first window daddy goes.
The plus changes to minus,
then baby is what shows.

The equal sign comes next,
then mommy's at the rear.
The third row is now finished,
our time to end is near.

3 4
1 = 4
3 = 4
1 = 3

On the fourth row,
daddy goes first once more.
The minus sign comes next,
then mommy we can't ignore.

Now comes the equal sign,
leaving baby very last.
We have finished every row.
Wow! We did that fast!

3 4
1 = 4
3 = 4
1 = 3
3 = 1

The Fact family's house is done.
Do you know their secrets yet?
There are some important keys
that you must not forget.

On the following pages,
these keys you will find,
You must always remember
to keep them in your mind.

3 4
1 = 4
3 = 4
1 = 3
3 = 1

If in a number sentence,
a plus sign you do find,
The big number we call daddy
is at the end assigned.

$$3 + 5 = 8$$

$$5 + 3 = 8$$

$$8 - 3 = 5$$

$$8 - 5 = 3$$

If in a number sentence,
you spot a minus sign,
Daddy must come first
in the number sentence line.

Mommy and little baby
 can switch places anytime.
Daddy can only move
 with the changing of the sign.

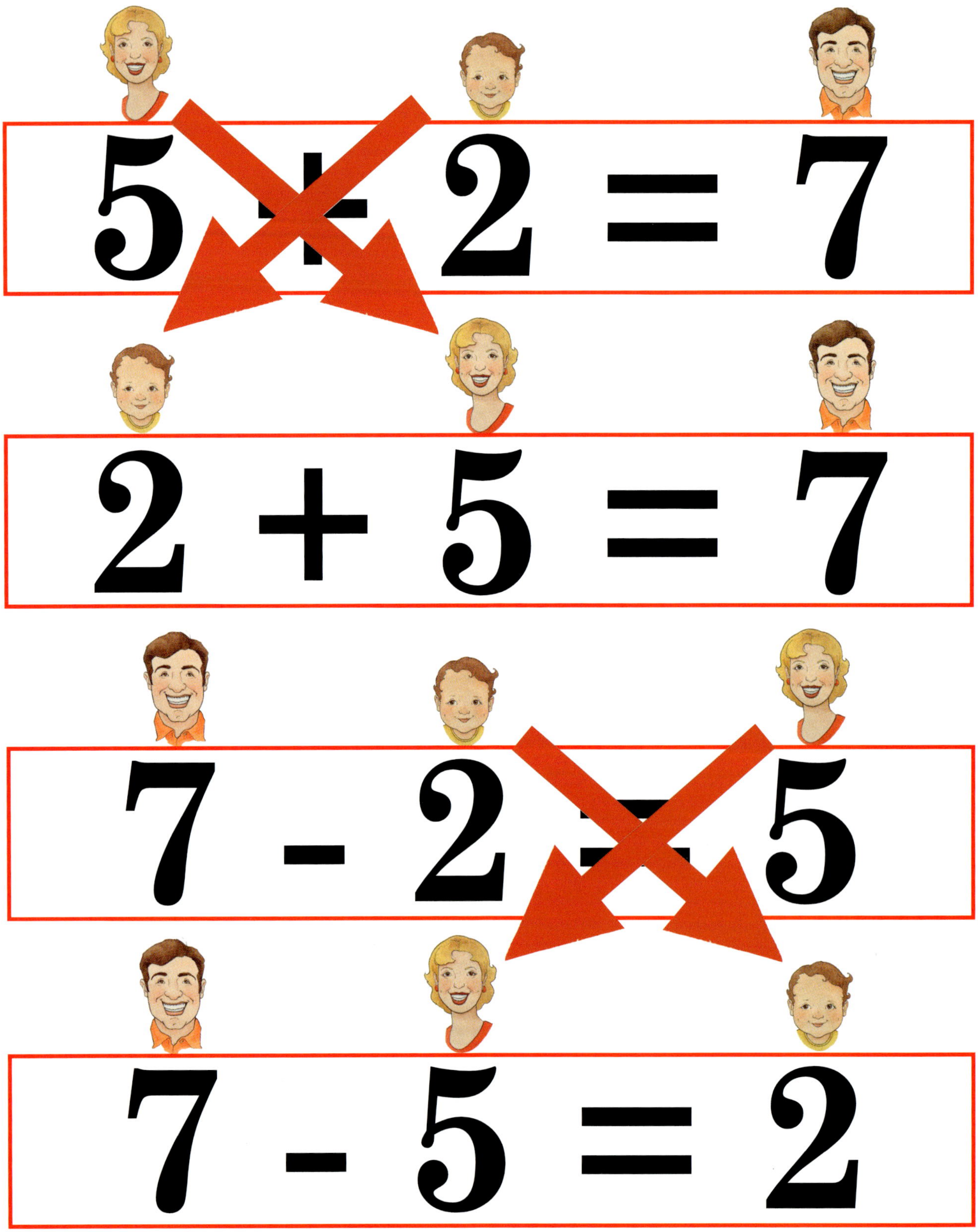

$$5 + 2 = 7$$

$$2 + 5 = 7$$

$$7 - 2 = 5$$

$$7 - 5 = 2$$

For one number sentence
to be related to another,
<u>All</u> the numbers being used,
must be the same as in the other.

So now you know the secrets
    the Fact family likes to share.
Of how they are related,
    you need to be aware.

$$9 + 3 = 12$$

$$3 + 9 = 12$$

$$12 - 3 = 9$$

$$12 - 9 = 3$$

With other number sentences,
    you must practice these great keys.
So understanding related facts,
    for you can be a breeze.